PROJECT VALUATION

AND

ESTIMATION OF WORK

ALEXANDER PAUL

Edition: 1

ISBN: 9798714316531

Author: ALEXANDER PAUL

TO THE READERS

This book deals with applications and analysis of important rules and techniques related to valuation and estimation of a construction projects. This book introduce you in the world of estimation and costing which is very helpful during the budgeting of your project such as residential or commercial building, penthouse, villa etc. estimation and valuation is key of success for any civil engineer. Quantity estimation is very important to survive in this competitive world for any civil engineer in whole world.

Contents

PROJECT ESTIMATION 5

QUANTITY MEASURMENT UNITS.. 7

ESTIMATION 8

TYPE OF AREA 11

TURNOUT OF WORK 12

VALUATION... 14

DEPRECIATION 19

UNIT CONVERSION... 21

Q & A 23

PROJECT ESTIMATION

Estimate – before starting the project we generally find the probable cost of project theoretically and use it as base.

Actual cost – it is the cost which determines after completion of project.

Project management – it is a skill by which we learn how to manage the entire project so that outcome is efficient.

There are three steps of project management.

1. Planning – to know objective of project.
2. Scheduling – allocation of resources and set the time for project.
3. Controlling – to reduce difference between scheduling and planning.

Detailed estimation – in this method of estimation we split the project in short work like earthwork, concrete work, electric work etc. and then calculate estimate and cost of each work.

Abstract of estimate – in this process we find cost formula which means the rate of item is multiplied by quantity of work.

Abstract = (Quantity of item) X (rate of item)

IMPORTANT TERMS IN ESTIMATION

Contingencies – it is the unforeseen expenditure which occurs during the life of project such as- Change in rates, change in design, accidents which directly affect the projects. It is generally taken 3 to 5% of total cost.

Work change establishment – it is indirectly connect to projects whose cost effect directly to estimate such as – supervisor, gatekeeper etc. it is taken 1.5% of total cost of project.

Water charge – it is generally taken as 1 % of cost.

Profit of contractor – it is taken as 10% for general case and 15% for small projects.

Electrification – it is around 8% of total cost of construction.

Sanitary work –it is also taken as 8% of total cost.

QUANTITY MEASURMENT UNITS

Here we are going to discuss of items of work and their unit measurement.

items	Unit of measurement
earthwork	Cubic meter (m^3)
Concrete work	M^3
soling	M^2
Damp proof course	M^2
r.c.c. work	M^3
shuttering	M^2
Glazing work	M^2
Cornice	M
lintel	M^3
Sanitary work	M
Jaffri work	M^2
skirting	M
Partition wall	M^2
electrifications	Nos.
bricks	Nos.
Brickwork 9"	M^3
cements	Bags
lime	Tons
Iron holdfast	Weight
bitumen	Tons
joints	M
Iron work	Kg/quintal
Plaster work	M^2

ESTIMATION

Types of estimation –

1. Plinth area estimation – plinth area is defined as built-up area or covered area at ground floor level or basement level. Excluding area of porch, mumty, internal shaft for sanitation installation whose area greater than 2 m². The calculation is taken from exterior wall to exterior wall in this type estimation.
2. Revised estimation – when original sanction is exceeded by 5% then estimation is prepared. And when the administration sanction is exceeded by 10% then revised estimation is prepared.
3. Supplementary estimate – when additional work is added then we prepare revised estimate of that work which is known as supplementary estimate.
4. Lump sum item estimate – some items which cannot be detailed and are small in quantity such as front architectural decoration, site

cleaning etc. these items are added in this type of estimation.

Analysis of rate – the determination of rate per unit item on the basis of cost of that items, requirement demand etc. is analysis. During costing, 10% contractor's profit added in items and if the transportation used to bring items is more than 8 km then transportation cost is also added separately. Analysis of rate depends upon –

I. Quantity of material
II. Availability of material
III. Quality of material
IV. Profit of the contractor
V. Type and no. of labours
VI. Water availability and transport charges.

Overhead cost – it includes taxes, rent, amenities of labour, stationary work, investments etc.

Overhead cost is of two types –

General overhead	Job overhead
Establishment	Interest of investments
Printing works	Amenities of labour
Rents, taxes	Supervision staffs

TYPE OF AREA

Plinth area – it is the built-up area/covered area.

Floor area – it is the total area of floor in between walls and it contains living hall, kitchen, staircase and 50% of balcony area.

Floor area is measured as interior of the wall to interior of wall.

FAR –the full form of FAR is floor area ratio and it is defined as the ratio of total area of all floors to total area of plot. It includes area of ground floor, first floor, second floor etc.

Circulation area – the area which allows movement is called circulation area. Examples are corridor, balcony, staircase etc.

Carpet area – carpet area is livable area. And it it calculated as –

Total floor area – circulation area.

Carpet area in residential building is 55 to 65% of total area.

Carpet area in commercial building is 60 to 75% of targeted area.

TURNOUT OF WORK

Turnout – turn-out of work is the output of work taken out for per person per day.

Type of work	Output/person/day
Brickwork in foundation and plinth	1.25 m^3
12 mm thick plaster	8 m^2
Cement concrete work (1:2:4)	5 m^3
Half brickwork/partition wall	5 m^2
Lime-concrete in foundation	8.5 m^3
Lime-concrete in roof	6 m^3
r.c.c. work	3.25 m^3
Distemper 1 coat	35 m^2
Excavation of soil	3 m^3

2.5 cm cement concrete floor	7.5 m^2

Some very important measurement of items –

Pillers measures in m^3

Honey comb structure measures in m^2

Brick edging measured in running meters

Thickness of walls measured in cm.

VALUATION

Schedule – list of items.

Schedule of rate – it is defined as the list of items used for projects to help us to decide estimate or rate of particular item are called schedule of rate.

Valuation – it is defined as a technique or method to determine priced value of assets such as building, roads etc.

Valuation of assets depends upon –

Locality, structure life, legal contract basis etc.

Purpose of valuation –

I. Buying and selling of property
II. Taxation
III. Rent fixation
IV. Security of loan or mortage
V. Compulsory acquisition
VI. Valuation of a property is also required for insurance, betterment charges and speculation etc.

Gross income = total income + outgoings + operational charges + collection charges.

Net income = gross income – outgoings.

Outgoings – it is the expenses which are required to be incurred to maintain the revenue of building.

Types of outgoings –

I. Taxes
II. Repairs
III. Management and collection charges.
IV. Sinking funds
V. Loss of rents
VI. Miscellaneous.

Scrap value – the cost of dismalted materials are called scrap value. Generally the scrap value of a building is about 10% of its total construction cost.

Salvage value – the value/cost of item at the end of the utility period i.e. after expiry date and without dismalting of material.

The cost of removal, sales etc. are not included in this value.

Market value – the market value is the cost of item at particular time in open market.

Note – salvage or scrap value may be negative, positive or can be zero.

Book value – the amount of item shown in account book is called book value.

Book value depends upon the amount of depreciation allows per year and it is reduced year by year and at the end of utility period of property and The remaining value will be only scrap value.

Sinking fund – a certain amount of money taken aside from rent by the owner to accumulate cost of construction are called sinking fund. The outgoings already includes in it.

Formulae –

$$I = \frac{S\,i}{x-1} \text{ where, i= annual installment}$$

required.

$$X = (1-i)^n$$

n = no. of years required to accumulate sinking fund.

S = total no. of sinking fund to be accumulated.

Capitalized value – it can be calculated by determining net income and highest prevailing rate of interest.

Capitalized value = net annual income X years of purchase.

Capital cost – total cost of construction + cost of land.

Obsolescence – the value of property decrease with time due to out of date design etc. this term is called obsolescence.

Annuity – it is a certain amount which is annual fixed, has to pay by contractor to his client.

Year's purchase – the capital sum required to invested in order to receive an annuity of 1 dollar at certain rate of interest.

Year's purchase = 100/i

I = rate of interest.

Year's of purchase in case of sinking fund –

$$100/(i+s)$$

S = sinking fund.

DEPRECIATION

The decrease in value of property due to structural deterioration, wear and tear, decay and obsolescence.

Methods of calculating depreciation –

Constant % method

Straight line method

Straight line method – here the loss of value of a property is constant at each year.

$$\text{Annual depreciation} = \frac{original\ cost - scrap\ value}{life\ of\ structure\ in\ years}$$

D=(C-S)/n

D=annual depreciation, c =original cost, S=scrap value, n=life of structure in years.

Constant % method – here the loss in value of property at constant %rate each year.

$$\text{Annual depreciation} = 1 - (S/C)^{1/n}$$

Value of property of depreciated cost at the end of first year = C_1=C-DC

Depreciation for second year=C_1-DC_2.when s=0this formula is fail.

Some other important unit measurement

- Tar and bitumen road generally measure in m^2
- Pre-mix carpet =m^2
- Grouting=m^2
- Cement concrete road =m^3
- Skirting and dado measure upto height of 30cm taken in m^2
- Filling of brickwork joints also called pointing .
- Pointing done by-
- Flush pointing
- Ruled pointing
- Weather and truck pointing

Construction cost in percentage-

Electrification – 8%

Sanitation – 8%

Contingencies – 2 to 5%

Work change establishment – 1.5 to 2 %

Overhead cost – 5 to 10%

Departmental charges – 10 to 15%

Tools – 1 to 1.5%

UNIT CONVERSION

Some important unit conversion in civil engineering

1 inch = 2.54 cm

12 inch = 1 foot

1 foot = 30 cm

1 meter = 3.28 foot

1 meter = 100 cm

10 mm = 1 cm

$1 m^2$ = 10.76 square foot

1 pound = 450 gram

$1 m^3$ = 35.31 cubic feet

Density of cement = 1440 kg/m^3

Density of sand = 1500 kg/m^3

Density of aggregate = 1500 – 1600 kg/m^3

Mostly we measure sand and aggregates in cubic feet and cement in bags where 1 bag of cement contain 50 kg cement.

Q & A

Q1. Find quantity of cement and sand in plaster work for 100 square feet area with mortar 1:4.

A1. So, the area = 100 ft^2

   Let the plaster work is for inner side hence we consider the thickness of plaster = 12 mm.

Now, area to be plastered in meter are = 100/10.76 = 9.3 m^2.

   So, volume = 9.3 X .012 = 0.112 m^3

   This is wet volume of plaster so, we have to convert it into dry volume hence multiply the value with 1.33.

   0.112 X 1.33 = 0.149 m^3

Here the mortar is 1:4 so,

Quantity of cement = 0.149 X (1/5) X 1440 = 42.9 kg, take 43 kg

Quantity of sand = 0.149 X (4/5) X 1500 = 178.8, say 179 kg

Quantity of sand in CFT = 0.149 X (4/5) x35.31 = 4.21 CFT.

Q2. What is setting out plan?

A2. Setting out plan use to set the boundary wall of total land so that to take the ownership of plot.

Q3. What is setback area?

A3. Set back area is non-tower area which is use for emergency purpose. It is taken on the basis of height of the tower or building. If the height of tower is 24 m than setback area is 9 meter. It may very in different part of country.

Q4. Calculate the quantity of tiles for 100 X 100 foot area?

A4. Area of floor in total = 10000 ft^2

If the area of 1 tile = 4 ft^2

Then, quantity of tiles = $\dfrac{10000}{4}$ = 2500 ft^2

If no. of tiles in one box = 5

So, no. of tile box required = 2500/5 = 100

If we consider rate of one box of tile =700

Total cost = 700 X 100 =70,000.

Q5. How to calculate weight of a hollow steel pipe having length 1 meter, inner diameter is 1 meter and outer diameter is 1.2 meter?

A5. Here, the volume of hollow pile is know as
$= (\pi R^2 - \pi r^2)\,L$

R=D/2

$v = \pi/4(D^2 - d^2)\,L$

$= 3.14/4(1.2^2 - 1^2)X\ 1= 0.345\ m^3$

Density of steel = 7850 kg/m^3

$= 0.345\ X\ 7850 = 2711.39$ kg.

Q6. What is floor area ratio?

A6. It is define as the ratio of total floor area to total plot area.

Q7. What is the final setting time of lime?

A7. It is about 16 to 18 hours.

Q8. Why we use ribbed/deformed bars in construction work?

Q8. We use ribbed bar to prevent slipping of bar from concrete and to increase bond strength between steel and concrete.

Q9.how much spacing provided between two ribs/deformation in a bar?

A9.it should not greater than $S < 0.7d$

Where s = spacing and d = diameter

Q10. Explain over reinforcement structure?

A10.in over reinforcement, chance of compression failure is high and no chance of tension failure and the failure of compression may be primary or secondary hence always remember never provide over reinforcement because sudden collapse of beam occur in over reinforcement.

Q11. Explain under reinforcement structure?

A11. In under reinforcement the chance of tension failure is high in comparison of compression failure

Hence sudden failure of structure not possible and it give initial hint of failure of structure and due to which the individual can secure himself within time.

Q12. What is the pH value of water used in concrete mixing?

A12. The pH value of water should not less than 6 and more than 9.

Q13. Can sea water used for mixing of concrete?

A13. The sea water not recommended for construction work.

Q14. How the cement bags stacked?

A14. Cement bags will stacked in shed which is dry in leak-proof and dry bricks lay in at least two layers to avoid direct contact of cement bags with floor.

Q15. What is the maximum silt content allow in sand?

A15. The maximum silt content allow is 8%.

Q16. How the manufacturing company supply the different form of lime?

A16. Quick lime always supply in lumps and the hydrated lime always supply in powered form and this is the easiest way to identify the type of lime.

Q17. How to specify the fine aggregates?

A17. The aggregate which is capable of passing 4.75 is sieve is categorized as fine aggregate and it is specified as coarse sand, fine sand, stone dust.

Coarse or fine sand be either river sand or pit sand or combination of both.

Q18. What is lug height?

A18. The lug or transverse height in a bar should not greater than 0.04d. Where d is the diameter of bar

Q19. What is the angle made by lugs and rib in bar?

A19. It is less than 45^0 always.

Q20. What is fe500 means?

A20. It means the yield strength of the bar is 500 N/mm^2.

Q21. How we provide extra bar in beam?

A21. We know that the maximum shear force in a beam is always acted at column junction because it is the compression zone and maximum bending moment in beam always at mid span because it is tension zone hence the chance of failure of beam is maximum at centre and to avoid the failure of structure we should provide top reinforcement in column junction and bottom reinforcement at mid span of beam.

Q22. What is the measuring equipment criteria of batching plant?

A22. The thumb rule of measuring critera of we can say the tolerance limit in material is –

Cement – ±2%

Aggregate – ±3%

Water = 3%

Mixing time in mixture = 2 minute.

Q23. What is construction bill?

A23. Construction bill are request or demand made by contractor from his Clint for his claims of his expenses for all work done by his team at site in writing. With all supporting documents such as measurement sheet, records, tax invoice with reference to agreed contract conditions.

Q24. How many types of construction bill?

A24. The construction bills depends upon the type of contract or agreement executed between clint and contractor. They are as follows-

I.   Cost plus contracts
II.   Time and material contracts
III.   %rate contract
IV.   Lump sum or fived rate contract
V.   Plinth area rate contract
VI.   Build up area rate contract
VII.   Prw (piece rate worker) contract
VIII. Labour rate contract
IX.   Epc (engineer procurement and construction)
X.   Trunky basis contract

Q25. What is ra bill?

A25. RA bill claimed on monthly basis or may be 15 days, 20 days basis as per agreement.

RA bill contains-

Request letter on company's letter head, tax invoice, BVS/MOP/COP(certificate of payment), measurement seat, bbs, reconciliation of steel and cement and Labour compliances such as PF AND ESI challans.

Q26. What is final bill?

A26. When work completed at site, then contractor submit his final bill in consolidated quantity i.e. for start to end or onwards last bills. And after submission of final bill, no claim will be considered.

Q27. What is NIT and EMD in construction work?

A27. NIT –notice inviting for tender and EMD means earnest money deposit which released in final bill

Q28. Named the type of tenders?

A28. Single tender, limited tender, open tender and Global tender.

Q29. What is budget?

A29. Budget is written approved document which ensure the amount of money is invested in future for prescribed project.

Q30. What is estimate?

A30. It is the part of research and development. It contains quantity survey as well as costing.

Q31. What is letter of intent?

A31.these letters are issued by Client to contractor. This letter issues after issue of tender, negotiation, rate fixation, technical aspects etc. now it's time to award of work and for which the client issues a letter to his contractor to start the project.

Q32. What is the mobilization period?

A32. The client introduce the mobilization period in letter of intent and this period may be of 15 days or one month.in this period the client give some fixed amount to his contractor to set up the basic site structure like store room, batching plant, site office, transit mixer, quality testing machine etc.

Q33. What is work order?

A33. The client issue a letter of award, also called work order after the completion of letter of intent and in this letter so many things included like article of agreement, BOQ, general contract condition (g.c.c.) or S.C.C. etc.

Q34. What is completion period?

A34. A fix time period to finish the due work.

Q35. What is liquidate damage?

A35. It is applied when any contractor/vender not fulfill his commitment of work within the completion period then certain amount of penalties applied on the contractor for delaying the project/work.

Q36. Explain 1-way slab?

A36. It is supported by beam in only two sides or direction and also follows the criteria of below equation.

$$\frac{longer\ span}{shorter\ span} > 2$$

Q37. Explain 2-way slab?

A37. The slab which is supported by beam in all four sides OR direction is termed as two way slab and also follow the criteria below equation.

$$\frac{longer\ span}{shorter\ span} \leq 2$$

Q38. Explain flat plate slab?

A38. The slab which is only supported on column or load bearing wall are termed as flat plate slab.

Q39. What are Cornish?

A39. It is like skirting but made near ceiling sides and generally made up of cement plaster.

Q40. How to calculate w/c ratio?

A40. W/C ratio can be vary from 0.45 to 0.6

If we consider 0.45 liter for 50kg of cement then .45 X 50kg = 22.5 liter water required for one bag of cement.

Q41. How to find quantity of paint for area of wall 11 X 60 feet?

A41. Here the area = 11 X60 = 660ft

We know than 1 gallon of UK =4.54litre

And one gallon of UK paint cover =350 square feet

So, 660/350 =2 gallon (approximately).

Q42. In residential building, in which direction kitchen is mostly preferred to installation?

A42. South-east

Q43.How much the opening of window given at-least with respect to floor area of room?

A43. Windows should be 15% or 1/8 of floor area.

And windows +doors=1/4 of floor area.

Q44. Find floor area ratio of building with total plinth area with total plinth area 550 m^2 with plot area is 34 X44 m

A44. FAR =(550)/(34X44)=0.3676

Q45. What is the minimum width to be provided for a staircase for educational building?

A45. 1.2m to 1.5m

Q46. How to calculate carpet area?

A46. Total floor area –circulation area.

Q47. What is the order of booking dimension in standard measurement book?

A47. L x B x H

Q48. What is nearest measurement of woodwork?

A48. 0.002m

Q49. What is the standard density of mild steel, copper, iron and aluminum?

A49. Mild steel -7850 kg/m^3

   Copper - 8840 -8940 kg/m^3

   Iron – 7580 -7720 kg/m^3

   Aluminum – 2640 – 2800 kg/m^3

Q50. The quantity of dry distemper required for single coat over 100 m^2?

A50. 6.5 KG for one coat.

Q51. Septic tank cannot be constructed by?

A51. Wood

Q52. Write descriptive information of septic tank?

A52. Septic tank should has depth= 1 -2m

        Length and width = 2 -4 m

        Detention time = 12 -24 hr.

Q53. Write the items of earthwork which is not measured separately?

A53. Items like Setting out of works, site clearance and dead men etc.

Q54. Find quantity of cement, sand, aggregate in 1 m^3 concrete with proportion 1:2:4

A54. So, grade of concrete – M15

Quantity of cement –

$$\frac{1.54 \times 1}{7} \times 1440 = 316.8 \text{ kg}$$

Quantity of Sand = $\frac{1.54 \times 2}{7} \times 35.31 = 15.54 \text{ ft}^3$

Quantity of aggregate = $\frac{1.54 \times 4}{7} \times 35.31 = 31.07 \text{ ft}^3$.

Q55. How many bricks use in 1 cubic meter?

A55. 500

www.ingramcontent.com/pod-product-compliance
Lightning Source LLC
Chambersburg PA
CBHW071121220526
45467CB00004B/1994